Book 9
On Rock

Contents

Introduction	On Rock	3
Chapter 1	Meet the Minerals	4
Chapter 2	Rocks of all Kinds	10
Chapter 3	How are they used?	16
Chapter 4	Technology Today — Making Minerals in a Lab	20
Check Up	Alberta Einstein's Brain Tester	22
Don't Gloss Over the Glossary!		24
Indexed!		25

Eric Einstein does an about-face at the rock face.

Heigh-ho! Heigh-ho! It's off rock hunting that I go!

Go easy with the rock hammer mate!

Introduction

On Rock

Do You Know...
- how to tell one mineral from another?
- how minerals are related to rocks?
- how rocks are grouped?
- how people use rocks and minerals?

Chapter 1   Meet the Minerals

■ How can you tell one mineral from another?

Look at the frog-shaped piece of jewellery below. It is made of three valuable minerals — gold, emeralds, and diamonds. Gold, emeralds, and diamonds are among the 2,000 or more minerals found in the earth. **Minerals** are solids that form from one or more chemicals in the earth. They are not living and do not come from living things.

Each mineral has several properties. Colour is one of those properties. Properties are used to identify minerals. Different minerals may have some of the same properties. For this reason, several properties are used to identify a mineral. Let's look at some properties used to identify minerals.

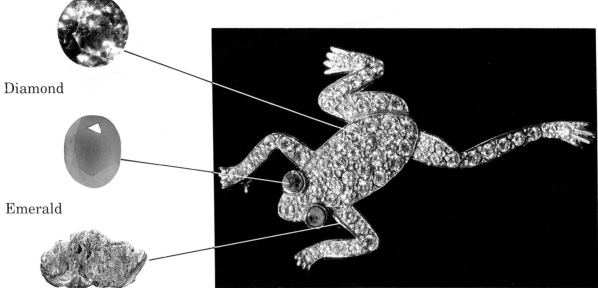

Diamond

Emerald

Gold

No warts on this frog!

Colour Minerals can be many colours. They can be bright red, yellow, green or blue. Often, colour is a clue to a mineral's identity. But colour should not be the only property used to identify a mineral. Look at the samples on the right. By looking only at colour, you might think that they are the same mineral. But they are different minerals. More than colour is needed to identify these minerals.

Azurite

Linarite

A mineral's colour can also change if it contains **impurities** (im-pyur-rih-tees). Pure minerals are made of certain chemicals. Any other chemical that joins with a pure mineral is an impurity. An impurity changes the colour of a mineral.

Look at the picture of three kinds of quartz. Only one is pure quartz. The other two have impurities.

Three kinds of quartz

Alberta Einstein plans to add to her rock collection.

Streak test for pyrite

Streak The surface colour of some minerals changes when they are exposed to air and weather for a long time. The colour you see is not the true colour. These minerals can best be identified by their streak colour.

Streak is the colour of a powdered mineral. To determine streak, a mineral is rubbed across a piece of special tile. The mineral makes a streak of coloured powder on the tile.

A mineral's streak is always the same colour. For example, pyrite (pie-rite) always leaves a greenish-black streak.

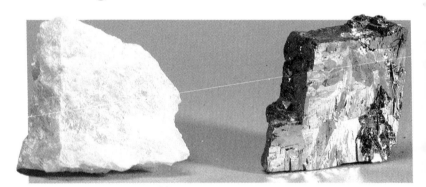

Talc (left) and galena

Pick five objects in your classroom. What word would you use to describe the lustre of each?

Lustre Some minerals reflect light and shine like polished metal. Other minerals are very dull. The way light reflects from a mineral's surface is called **lustre**.

Look at the picture of the talc and galena (gah-lee-nah) above. Compare their lustre. Which of these minerals shines like metal? Minerals that do not shine like metal may be described as dull, pearly, earthy, or silky.

Hardness Minerals can be hard or soft. This property is known as **hardness**. When rubbed together, a harder mineral will scratch a softer one. Often, minerals are identified by their hardness.

Hardness is measured on a scale of 1 to 10. Diamond is the hardest mineral, with a value of 10. A diamond can be scratched only by another diamond.

Look at Mohs' hardness scale. It uses ten minerals to test for hardness. Some common objects that can be used are also listed.

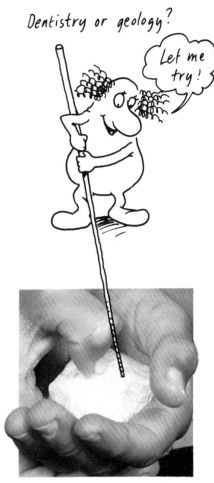

Fingernail scratching talc

Mohs' Scale of Hardness		
Mineral	Softest	Common Objects
1 Talc		
2 Gypsum		
		Your fingernail ($2\frac{1}{2}$)
3 Calcite		
		A copper penny ($3\frac{1}{2}$)
4 Fluorite		
5 Apatite		
		Glass or iron nail ($5\frac{1}{2}$)
6 Orthoclase feldspar		
7 Quartz		
8 Topaz		
9 Ruby or Sapphire		
10 Diamond		
	Hardest	

Penny scratching calcite

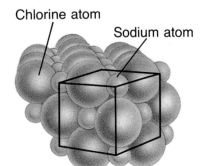

Pattern of atoms in a salt crystal

Salt crystals

Crystal Shape The building blocks, or atoms, that make up minerals are arranged in patterns. Each mineral has its own special pattern. This arrangement of atoms in minerals forms **crystals** (kriss-tils). Since each mineral has its own crystal shape, crystals can be used to identify minerals.

Look at the picture of salt crystals. You can see that these salt crystals are shaped like cubes. The cube-shaped crystals are caused by the arrangement of atoms in salt.

Now look at the picture of the amethyst (am-ih-thist) crystals. The atoms that make up amethyst are not arranged in the same pattern as the atoms found in salt. The result is a different-shaped crystal.

The size of a mineral crystal depends on how slowly the crystal forms. Large crystals take more time to form. No matter what the size is, the shape will always be the same.

Alberta Einstein dreams of an amethyst ring.

Amethyst crystals

The Greed for Precious Minerals

The tale of the Koh-i-noor diamond.

Once upon a real time there was a beautiful 108 carat diamond. It was owned by a royal Indian family. They gave it to a conquering Mogul prince because of the many kindnesses he had shown to the people he had over-run.

In 1526, the diamond was given by Prince Humayun to his father — the new Mogul Emperor of India, Babur. But Babur later returned it to his son. This son became the second Emperor of India.

In 1739, the Shah of Iran rode into Delhi and seized the precious diamond for himself, but it was later reclaimed for India by the Sikhs.

Crown of Queen Elizabeth (The Queen Mother). The Koh-i-noor is the large, roundish diamond at the front.

The tale does not end here. In 1849 the diamond ended up in the hands of the English after they had seized the land from the Sikhs. It was taken to England where Queen Victoria had it re-cut and wore it as an ornament. It was the central stone in the crowns of Queen Mary and Queen Elizabeth (the Queen Mother) at their coronations in 1911 and 1937. The Koh-i-noor diamond has not been worn by a King because it is supposed to bring bad luck to any man who wears it. Today it is displayed in the Tower of London with the British Crown Jewels. Perhaps one day it will be returned to its original owners. What do you think?

Babur

Chapter 2 Rocks of all Kinds

■ How can all rocks be sorted into three large groups?

You are never far from rocks even if you live in a city. There are little rocks that get in your shoes. There are big rocks that can be seen where a road has been cut. On the surface and deep underground, rocks are everywhere.

Look at the picture of granite (gran-it). You may have seen this rock before. You can see that it is speckled with different colours. These speckles are small pieces of different minerals. They combine to make granite. In fact, all rocks are mixtures of minerals.

Mica

Quartz

Feldspar

Granite

People who study rocks and minerals are called **geologists** (jee-ol-oh-jists). They place all rocks in three groups: igneous, sedimentary, and metamorphic. Let's find out how the three groups of rocks are different.

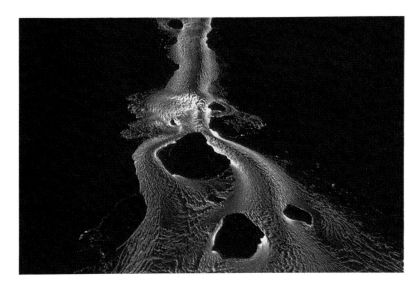
Melted rock

A rock band.

Igneous Rocks Deep inside the earth, it is so hot that some rock melts. This melted rock is called **magma**. If magma reaches the earth's surface, it is called **lava**.

As magma and lava cool, they usually form mineral crystals. These crystals join to form rock. Rocks that form from the cooling of magma or lava are called **igneous rocks** (ig-nee-iss).

Crystals can form quickly or very slowly. Crystals that form slowly have more time to grow. They are larger than those that form quickly.

Basalt (bass-solt) and granite are common igneous rocks. Look at the picture of basalt. Compare it to granite. Granite formed underground from magma. Basalt formed from lava, which cooled above ground. Why does granite have larger crystals than basalt?

A block of basalt

ALBERTA NEEDS YOUR HELP WITH THIS...
There is no atmosphere on the moon. What type of rock would you not expect to find there?

Think Zone

THINK THINK THINK THINK THINK THINK THINK THINK THINK

Sedimentary Rocks Wind and water change the earth's surface. They can break large rocks into small pieces called sediments. The sediments can be caried away by wind or running water and deposited in layers that become very deep. Rocks that form from layers of sediment are called **sedimentary rocks** (sed-ih-men-tur-ree).

The pictures show one way that sedimentary rocks can form. The kind of sedimentary rock that forms depends on the kind of sediment. Sandstone is made from grains of sand. It feels gritty. Shale is pressed mud. It feels smooth. Limestone can be made up of small pieces of animal shells. How would it feel?

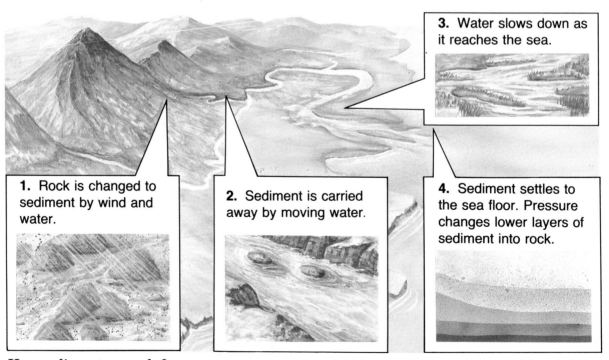

1. Rock is changed to sediment by wind and water.

2. Sediment is carried away by moving water.

3. Water slows down as it reaches the sea.

4. Sediment settles to the sea floor. Pressure changes lower layers of sediment into rock.

How sedimentary rock forms

Changing, Changing, Changing . . .

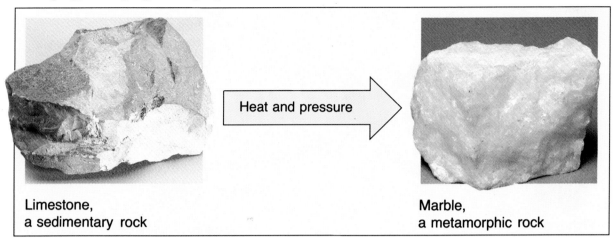

How metamorphic rock forms

Metamorphic Rocks Heat within the earth can melt rock, forming magma and lava. You might be surprised to know that heat can also change rocks from one kind to another. Pressure can change rocks from one kind to another, too.

Some igneous and sedimentary rocks are buried deep in the earth. Here, the pressure is very great, and the temperature is very hot. The extreme heat and pressure can change the minerals that make up a rock. Rocks formed when the minerals are changed by heat and pressure are called **metamorphic rocks**.

Marble is a metamorphic rock. Before it was changed by heat and pressure, it was limestone, a sedimentary rock. Slate, used to make chalkboards, is also a metamorphic rock. It was sedimentary shale before heat and pressure changed it to slate.

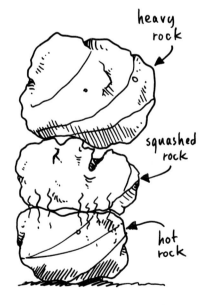

Rock undergoing trial by heat and pressure.

13

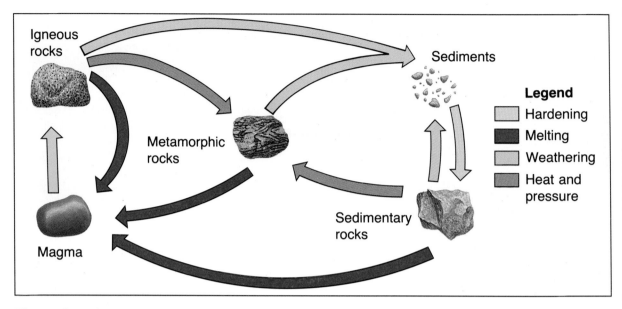

The rock cycle

You wouldn't want to wait around to watch a lump of magma grow up!

Rock Cycle A cycle is something that repeats itself. Seasons follow a cycle. Even life itself goes through a cycle. But did you know that rocks go through a cycle?

Rocks change from one kind to another. The change usually takes place over millions of years. It is certainly not the kind of change you can watch. This changing of rocks from one kind to another is called the **rock cycle**.

The drawing of the rock cycle shows that rocks change from one group to another. It also shows that heat and pressure and weathering cause rocks to change. Rocks are constantly forming, then changing.

You complete a cycle when you start at one place and come back to the same place. Start any place in the rock cycle and come back to the same place. Notice each change.

This is one ancient rock that's not for collecting . . .

. . . Uluru — Australia's most famous rock, 'born' in an ancient time and destined to last for an eternity.

Chapter 3   How are they used?

■ How do we use rocks and minerals?

We use minerals to manufacture almost everything — including aeroplanes, cars, paint, and batteries. We use material from rocks to build houses, roads, and schools.

In your home, you could find minerals in medicines, foods, dishes, and products used to clean the house. Even a common pencil has several minerals in it.

What is made of rocks and minerals?

Rocks and Minerals in Buildings

Skyscrapers are a common sight in cities around the world. These huge buildings poking into the sky could not be built without rocks and minerals.

Minerals found in steel.

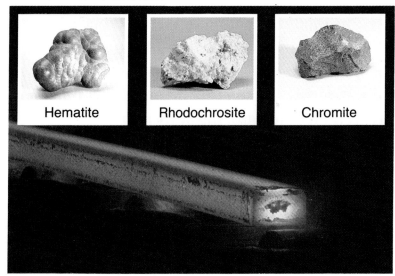

Meanwhile, back at the foundry . . .

First, a strong foundation is needed to support a skyscraper. Concrete foundations are made from sand, cement, and gravel. But concrete is not enough. It is made stronger with steel rods. Steel is made from iron and other minerals. So, without rocks and minerals we could not even begin to build a skyscraper.

Metals such as copper, aluminium, and steel are needed, too. Steel beams form the building's "skeleton". Copper wires will carry electricity. Copper pipes will carry water.

Many skyscrapers are covered with walls of glass. It may be hard to believe, but glass is made from minerals found in sand.

Granite, marble, and sandstone are types of rocks used in skyscrapers. The next time you see a skyscraper, look at it closely. Try to find ways rocks and minerals are used.

Rock in the city

17

Minerals in Technology Minerals are used in modern technology. In a laboratory, a material from quartz is used to make silicon crystals. The crystals are used to make chips for computers. The silicon chip has a pattern printed on it for storing information.

Quartz crystals like the one shown have become common in watches and clocks. An electric current can make the crystals vibrate at a very steady rate. Small batteries supply the current. After a certain number of vibrations, the hands tick off one minute.

Using quartz crystals helps us keep very accurate time. A watch with a quartz crystal may lose less than one second a year!

Silicon chips keep a computer's 'heart' beating . . .

Another Kind of Roc ...

This is one roc you won't find on your next rock hunting trip — well, I hope not anyway! If you're unfortunate enough to meet one it would probably carry you off to its eyrie high in a mountain and tear you limb from limb. This kind of roc is really a gaint mythological birdie-come-beastie. It was so strong, it could carry an elephant in its talons, and once a sailor called Sinbad tied himself to a roc's leg ... but that's another story ...

Chapter 4   Technology Today

Synthetic sapphire under special light

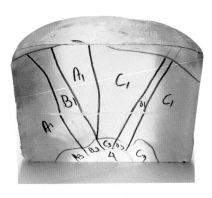

Synthetic sapphire before cutting

eyes like sapphires

Making Minerals in a Lab

In 1837, a scientist made a ruby in a lab. Since then, scientists have been able to make many other synthetic (sin-theh-tik) minerals. "Synthetic" means made by people. These minerals look a lot like the real things. It takes an expert to tell the difference.

Scientists can now make rubies, emeralds, sapphires, and diamonds in a lab. However, it takes a lot of heat and pressure.

Synthetic rubies, emeralds, sapphires, and diamonds are just as hard as the natural minerals. However, they are less expensive. Synthetic minerals can be made in a few minutes. Natural minerals may take years to form. Also, natural minerals are rare.

Some synthetic minerals are large enough to use as gems. Synthetic diamonds, however, are tiny, but they are widely used. Diamond-tip drills can easily cut through hard substances. Some needles on record players have diamond tips. They last a long time.

Think About It

Some people will buy only natural gems, even though they are more expensive. Why do you think they do this? Would you pay the price for a natural gem? Why or why not?

How many of these gemstones do you recognise?

ANSWERS: 1 Emerald **2** Sapphire **3** Diamond **4** Amethyst **5** Beryl **6** Garnet **7** Topaz **8** Black opal **9** Moonstone. You will notice that some of these stones are "rough" and some are "cut". Cut means that they have been cut and polished to be made into jewellery and other items.

21

Check Up   Alberta Einstein's Brain Tester

"Eric, you need a brain check-up! Heh! Heh!"

Summary
- Colour, streak, lustre, hardness, and crystal shape are properties used to identify minerals.
- Rocks are made up of minerals.
- Rocks can be sorted into three groups: igneous, sedimentary, and metamorphic.
- People use rocks and minerals in buildings, in technology, and in many other ways.

Science Ideas

1. What does the picture on the left show? How does it relate to rocks and minerals.

2. Make a list from **a** to **d**. Write whether each statement is fact or opinion.
   a. Rocks are made of minerals.
   b. Quartz clocks are the best clocks.
   c. Minerals with lustre are prettier than dull minerals.
   d. Rocks change from one kind to another.

Minerals in Australia

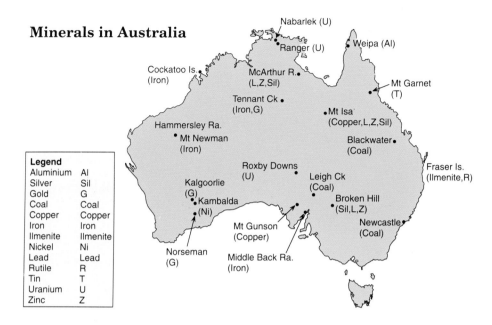

3. **Data Bank**

    Use the map to answer the following questions.
    1. What are five minerals found in Australia?
    2. Where are the most gold sources?
    3. Which mineral is found in the most places?

4. **Problem Solving**

    A certain mineral is needed to produce a very important material, which your country uses. Another country of the world is the only one known to have deposits of the mineral you need. Imagine you have been put in charge of national mineral supplies for your country. What would you do, and why?

Eric thinks of his answer

23

Don't Gloss Over the Glossary!

common — used everyday/something you would know
crystals — the arrangement of atoms in minerals
determine — to find out/discover
exposed — for all to see/in the open air/uncovered
formed — was made
foundation — a base
geologist — a person who studies rocks and minerals
hardness — a property of minerals often used for identification
igneous rock — a rock that forms when lava or magma cools down
impurities — markings that stop something from being pure/blemishes
lava — melted rock that gushes from volcanoes
lustre — the way light reflects from the surface of something
magma — melted rock from deep within the earth
metamorphic rock — a rock formed when minerals are changed by heat and pressure
minerals — a solid that forms from one or more chemicals in the earth/usually found combined in rocks
rock cycle — the changing of rocks from one kind to another
sedimentary rock — a rock that forms from layers of sediment

Alberta Einstein is spooked out by the spirits of the rocks.

Book 4
We Need Each Other

Contents

Introduction	We Need Each Other	3
Chapter 1	All for the Sake of Food	4
Chapter 2	Depend on the Decomposers	10
Chapter 3	Other Things to Depend On	14
Chapter 4	Technology Today — Zoodoo For Sale	20
Check Up	Alberta Einstein's Brain Tester	22
Don't Gloss Over the Glossary!		24
Indexed!		25

Eric Einstein tries to teach his vulture table manners, and Alberta Einstein gives up acting.

Introduction

We Need Each Other

Do You Know...
- how plants and animals depend on each other for food?
- the importance of decomposers?
- how plants and animals depend on each other for shelter, protection, and reproduction?

Chapter 1   All for the Sake of Food

■ What are some different ways in which animals relate to each other in their search for food?

The earth is home for many living things. They share the land, air, and water. To survive, each living thing depends on others. Life on the earth is a web of relationships among living things.

The way living things, or organisms, relate depends often on their need for food. Look at the picture. The way a brown bear relates to a fish is affected by the bear's need for food.

Sometimes, nature's relationships seem cruel. Some relationships between two organisms do not harm either organism. But other relationships can harm one of the organisms. In this chapter, you will learn how some plants and animals depend on other plants and animals for food.

Eric Einstein tries to teach a wild bear to fetch fish.

This bear depends on fish for food, not Eric . . .

Scavengers Some consumers do not capture and kill their food. They get their food from plants and animals that have died. Animals that eat dead plants and animals are called **scavengers** (skav-in-jerz).

The pictures show three scavengers. Each one spends time searching for animals that have died. Vultures can often be seen soaring above the remains of a dead animal. When it is safe, they land and begin picking away at the leftovers. Hyenas live in Africa. After a lion has killed an animal and eaten its fill, the hyenas eat the leftovers. The crab does its work on the ocean floor. Dead fish serve as the crab's food.

Scavengers are an important link in the food chain. They clean up food left behind by other consumers. Scavengers are usually one of the last links in a food chain.

Vultures are scavengers.

Crabs are scavengers.

Hyenas are scavengers.

Is Eric Einstein a scavenger?

Fleas on the hop looking for a new host.

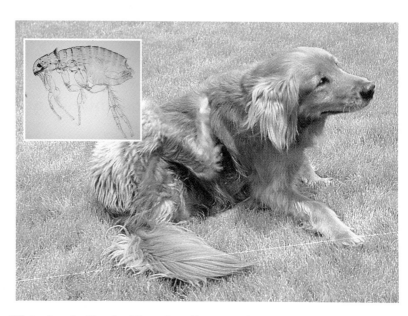
This dog is tired of hosting flea parties.

Parasites If your dog or cat is always scratching, it probably has fleas. Fleas get food by sucking other animals' blood. Organisms that live on or in other organisms and do harm to them are called **parasites** (par-a-sites). They get food from the organism they live on or in. The organism they live on or in is called a **host**. Fleas are parasites. Dogs and cats are hosts.

Some of the most common parasites are types of worms. Tapeworms live inside the body of another animal. They take food their host has digested. Leeches are worms that live outside the body of another animal. Leeches suck blood from the host animal.

Even some plants are parasites. Mistletoe is a parasite that grows on trees. It takes food and water from the tree it lives on.

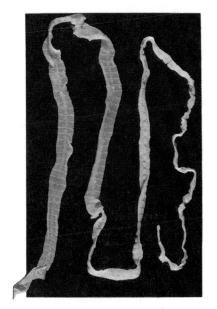
Tapeworm pretending to be a Christmas ribbon . . .

Mutualism Some animals have a special relationship with each other called mutualism. In **mutualism,** each organism does something that benefits the other organism.

The clownfish and the sea anemone (uh-nem-in-nee) in the picture are examples. The anemone can sting most fish. But it does not sting the clownfish. The anemone helps keep the clownfish safe from enemies. At the same time, any food the clownfish drops becomes food for the anemone.

Mutualism is massaging your friend's feet.

Clownfish and anemones help each other.

A **lichen** (lie-ken) is a plantlike organism that is really two organisms. One of the organisms is an alga. The alga makes its own food, which it shares with the other organism, a fungus. The fungus cannot make food, but it can take in water, which it shares with the alga. Since these organisms help each other, a lichen is a good example of mutualism.

Lichens lounging about

ALBERTA NEEDS YOUR HELP WITH THIS:
"Commensalism" comes from two Latin words that mean "together" and "table". How do those words relate to our meaning of "commensalism"?

Think Zone

THINK THINK THINK THINK THINK THINK THINK THINK THINK THINK

Commensalism With some organisms that live together, only one of the organisms benefits. But the other organism is not harmed. This special relationship is called **commensalism** (kom-men-sil-iz-im).

Look at the picture of the whale. The white patches on the whale are really other organisms, called barnacles (bar-na-kls). Barnacles are animals that have hard shells. Barnacles also grow on rocks, boats, and other objects found in salt water. The barnacles growing on the whale do not harm the whale. The barnacles benefit by eating food stirred up by the whale.

Barnacles hitching a ride on a whale.

(A fun and seriously-silly questionnaire)

Could you be found guilty of scavenging, parasitism, mutualism, or commensalism?

1. When you cut a cabbage from the garden, would you . . .
   a pronounce it dead and eat it
   b put it out for other two-legged scavengers
   c feel guilty and give it a decent burial

2. When you finish school, you plan to
   a stay at home with Mum and Dad for free, forever and ever and ever
   b worm your way into someone's heart and marry them only if they're very rich
   c become a vampire
   d none of the above

3. You very much like to . . .
   a repay kindnesses done to you
   b brush your dog's coat if he or she will take you for your daily run
   c fend for yourself and expect everyone else to do the same.

4. Mutualism is when . . .
   a you go to the dentist and he'll clean your teeth if you pay him to
   b you help an old lady across the road
   c you take a partner and massage each others' feet
   d you share your lunch with a fungus

5. You have a great time holidaying with your auntie but you drive her crazy with your loud rap music so she buys herself some ear plugs. Is that . . .
   a scavenging   d commensalism
   b parasitism   e all the above
   c mutualism   f none of the above

6. You keep a boa constrictor for a pet. You feed it, take it for walk, let it sleep in your bed and knit it a long scarf for winter. Would you say your relationship was . . .
   a wonderful for the snake
   b harmful for the snake
   c great for the snake but not much use to you
   d great for you but boring and useless for the snake

7. If you could be anything in the world, you'd choose to . . .
   a be a tapeworm
   b host a flea circus
   c be a hitch-hiker in a whale's galaxy
   d develop a life-long relationship with an anemone
   e be a rubbish dump hermit

(Check the score on page 25)

9

Chapter 2   Depend on the Decomposers

■ Why are decomposers needed to complete the recycling of nutrients in nature?

In a food chain, plants are producers. They produce, or make, their own food. To make food, plants get nutrients from soil. These nutrients enter plants in the water taken in by the roots. But the nutrients plants remove from soil must be replaced. Only then can plants continue to grow in the soil.

Putting nutrients back into soil is the work of decomposers (dee-kom-poh-zurs). **Decomposers** are organisms that break down dead plants and animals. Decomposers are important because they replace materials that were once removed from the soil by plants.

Alas! Poor Yorick! I know you'll be decomposing well...

Alberta Einstein ponders the cycle of life .... and how the soil will be happy.

Grey whale beginning to decompose on beach

Types of Decomposers Tiny organisms called bacteria are one type of decomposer. Bacteria are so small that it would take more than one thousand of them to cover the full stop at the end of this sentence. Bacteria cause meat to rot and milk to sour.

Mushrooms are another type of decomposer. Mushrooms cannot make food. They get food from dead plants. As they do this, they break down dead plant parts.

Moulds are another type of decomposer. Like mushrooms, moulds cannot make food. They get food from other things. You may have seen green mould, which grows on cheese. You may also have seen black mould or blue mould, which grows on bread. Moulds break up, or decompose, dead organisms and waste matter.

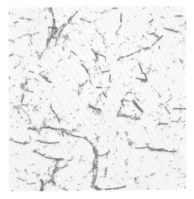

Bacteria like these are decomposers.

Many organisms helped to decompose this cactus.

A mouldy character.

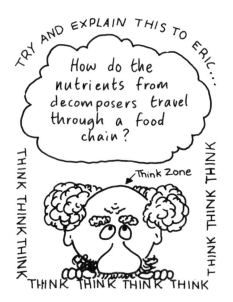

Some worms and insects help decompose materials. For example, flies lay eggs on the bodies of dead animals. These eggs hatch into larvae (lar-vay) called maggots. Maggots break down the bodies of dead organisms. In this way, fly maggots are decomposers.

Earthworms eat dead plant parts. They digest, or break down, the plant parts and then release them back into the soil. These digested plant parts add nutrients to the soil.

Look at the drawing to see how materials from plants and animals form a cycle. Follow the numbers to learn how nutrients move into and out of the soil.

The Decomposers' Dance . . .

Out of death comes life

Meet the Tumblebug

Someone's got to be the one to do the dirty work, but what a marvellous job they do. The Tumblebug (alias the dung beetle) is one such character. It uses its head and antennae to roll dung into huge balls (some of them more than six times bigger than its own body) which it then buries and feeds on. The dung also supplies instant 'takeaway' food for its young when they hatch from their smelly home.

In ancient Egypt, the scarab dung-rolling beetle was considered sacred. Many ancient pieces of Egyptian jewellery were made in the shape of scarabs.

Chapter 3 Other Things to Depend On

■ What else do plants and animals depend on each other for, besides food?

In order to stay alive, all plants and animals need food, shelter, and protection. They also have to reproduce so that new plants and animals can replace those that die.

In this chapter, you will find out how some living things use other living things for shelter and protection. You will also learn how some plants and animals depend on other plants and animals in order to reproduce.

Home for a hermit crab is an empty snail shell

This owl is as snug as a bug in a cactus rug.

Shelter and Protection Many animals depend on plants for shelter and nesting sites. These shelters protect animals from bad weather, and sometimes from enemies.

For example, the elf owl makes its home in a cactus. It is a safe place to lay eggs and raise young. In the cactus, snakes cannot get to the eggs or young owls.

Animals often depend on other animals for shelter. Some snakes, for example, live in holes in the ground. However, they cannot dig, so they take over burrows dug by other animals. In the USA, bluebirds live in holes in dead trees, but they cannot make the holes. They take over holes made by flickers and woodpeckers. (In Australia there are no birds that can make their own holes.) The hermit crab cannot grow a shell for protection. It takes over the shell of a sea snail.

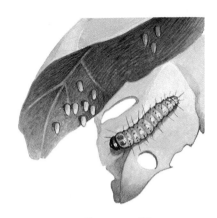

One small caterpillar who's got it all — food, protection, shelter...

Some insects lay their eggs on the undersides of leaves. There, the eggs are protected from rain and from the hot sun. Also, the eggs cannot be seen easily by other animals. The insect larvae that hatch from the eggs eat the leaves. The leaves provide both protection and food.

New tenants in town!

TRY AND EXPLAIN THIS TO ERIC...
Why is it important for seeds to scatter away from the parent plant that produced them?

THINK THINK THINK THINK THINK THINK THINK THINK THINK

Reproduction You might think that plants would be better off without animals. After all, it seems many animals just go around eating plants. But many plants could not make seeds without the help of animals. Seeds are the way many plants reproduce. For this reason, the survival of some kinds of seed plants depends on animals.

Bees, butterflies, and birds such as honeyeaters and spinebills, spend a lot of time visiting flowers. They remove food called nectar from the flower. As they are removing nectar, they pick up pollen. Pollen is a yellow powder made by the male part of a flower. The pollen is moved by the animals to the female part of a flower. The female part of a flower uses pollen to make seeds. In time, the seed will become a new plant.

Birds such as this Eastern Spinebill help plants reproduce.

Some plants also depend on animals to spread seeds. The dog in the picture has rubbed against seeds that stick to its fur. It will carry the seeds around until they fall off. Birds like the emu eat fruit. These birds often eat the seeds of the fruit, too. The seeds usually are not digested. The birds get rid of the seeds with their other wastes. The seeds often grow where the waste is dropped.

There are even animals that depend on other animals to help them give birth to their young. The female cuckoo lays her eggs in other birds' nests. She finds nests with eggs in them and lays one egg in each nest. The cuckoo egg hatches along with the other eggs. When the cuckoo egg hatches, the baby cuckoo tips the other babies out of the nest.

Dog carrying plant seeds

Cuckoo, cuckoo, cuckoo . . .

Emus on Arbor Day

Alberta gets excited

How People Use Living Things People use plants for food, shelter, clothing, and medicine. Plants make oxygen, which people need to breathe. Plants also decorate homes, gardens, and parks.

People use animals for food and clothing. Animals such as horses, camels, and donkeys do work for people. Animals kept as pets make many people happy.

Think of all the ways plants and animals make life better for you. It is important to care for the living things around you.

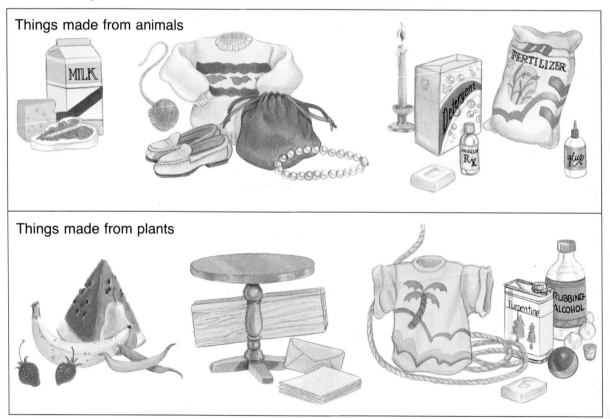

Where would we be without them?

Tale of a Pipe and an Adaptable Crab in Paradise

Once upon a summer time there was a man who went to the Maldives for a holiday. All day long he splashed and played in the warm waters of the lagoon. And in the evenings he would take out his pipe and puff away while he watched the sun go down.

Man at ease by seas.

Man sending out a distress call

Day in and day out it was always the same. And wherever the man went, so did his pipe. But one terrible day, the man arrived back at his hut just as the sun was setting and could not find his pipe anywhere. He hunted high and low all over the island. 'What happened to it?' he cried to the gods of paradise.

Suddenly the answer flashed to him with the last rays of the setting sun. His pipe must have fallen from his backpack as he had waded homewards through the shimmering, warm waters of the lagoon. Yes, it lay where it had fallen, cushioned in the lips of sand.

A school of pipe fish?

Stingrays sauntered lazily over it, schools of tiny inquisitive fish nudged it and a homeless hermit crab found it. The pipe bowl fitted snugly over its back making a perfect protective shell. And visitors to paradise marvelled at the peculiar sight of a pipe silently propelling itself across the soft silky sands on the lagoon floor.

Coolest crab in paradise.

Chapter 4   Technology Today

Lion

Rhinoceros

ZooDoo For Sale

Every day, animals at the San Francisco Zoo produce a tonne of waste. The waste used to be taken to a dump. Then someone had an idea. Decomposers could break the waste down into nutrients that plants need. Why not recycle the zoo waste as fertilizer?

Fertilizer is a material that is added to soil. It contains nutrients that make plants grow better. Recycling the zoo wastes as fertilizer would save a lot of money.

A recycling centre was set up near the zoo. Each day's waste is brought there. It is piled up and covered with straw. Air is pumped through the piles. They are also watered. After three months, the decomposed waste is dried. Then it is packaged for sale. The fertilizer is called ZooDoo.

Packaged ZooDoo

Think About It

People buy ZooDoo to loosen soil and fertilize plants. They say flowers and vegetables grow better with ZooDoo than with other fertilizers. Would you buy ZooDoo?

Should more zoos make ZooDoo? Why do you think they do not? What else could zoos recycle? Could the zoo's idea for recycling be used by others? Explain.

What's New At Your Zoo?

Many of the world's countries recycle their waste. Find out if recycling is happening at your nearest zoo, or anywhere in your local area.

Check Up   Alberta Einstein's Brain Tester

> **Summary**
> - Living things share the earth's land, air, and water.
> - Living things depend on each other for food, shelter, protection, and reproduction.
> - Decomposers break down dead plants and animals and put nutrients back in the soil.
> - People use plants and animals for food, shelter, clothing, and oxygen.

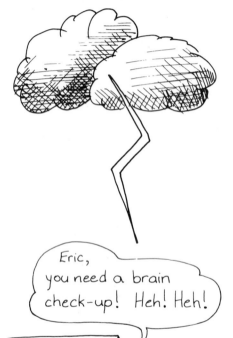

Eric, you need a brain check-up! Heh! Heh!

Science Ideas

1. Make a list from **a** to **c**. Write the statement that tells what is happening in each drawing.

 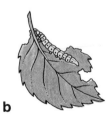 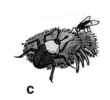

a   b   c

Consumer eats producer.
Decomposer breaks down wastes into soil.
Producer gets some nutrients from soil.

2. Make a list from **a** to **c**. Write the correct word to answer each question.
   a. What are organisms that live on or in other organisms and do them harm?
   b. What is the relationship in which only one of the organisms is helped?
   c. What are animals called that eat dead animals or leftover food?

3. Look at the picture. What animals can you think of that help each other?

4. Can you put the following words into a sentence each, that will show their meaning?

   scavenger   decompose
   host   parasite

5. **Problem Solving**
   The fallen river gum on the edge of Victoria State Forest is your favourite seat for observing the beauty of nature. However, it is no longer strong enough to hold you. The tree is hollow and filled with small bits of wood and many tiny worms. What is happening to the tree? What can you do to see that there will always be trees in Victoria State Forest?

The Thinker

Don't Gloss Over the Glossary!

algae — a plant like seaweed that can be found in fresh or seawater/ sometimes tiny and sometimes very long with trunk-like stems

commensalism — a relationship between two organisms in which one organism benefits and the other is not affected

decomposers — an organism that breaks down dead plants and animals

host — the organism on or in which an organism lives

larvae — the young of an insect

lichen — a plant-like organism that is made up of an alga which makes food and a fungus which takes in and stores water

mutualism — a relationship between two organisms in which each organism does something that benefits the other

nectar — sweet sticky liquid found in plants which attracts insects and birds that pollinate the flower

parasites — an organism that lives on or in another organism and does harm to it

scavengers — an animal that eats dead plants and animals

Alberta Einstein wonders if she is a scavenger....